DEPARTMENT OF COMMERCE

TECHNOLOGIC PAPERS

OF THE

BUREAU OF STANDARDS

S. W. STRATTON, DIRECTOR

No. 138

EFFECTS OF GLUCOSE AND SALTS ON THE
WEARING QUALITY OF SOLE LEATHER

BY

P. L. WORMELEY, Associate Physicist

R. C. BOWKER, Assistant Mechanical Engineer

R. W. HART, Assistant Physicist

L. M. WHITMORE, Assistant Chemist
Bureau of Standards

IN COOPERATION WITH

J. B. CHURCHILL, Director American Leather
Research Laboratory

ISSUED OCTOBER 6, 1919

I0111178

A History of Shoemaking

Shoemaking, at its simplest, is the process of making footwear. Whilst the art has now been largely superseded by mass-volume industrial production, for most of history, making shoes was an individual, artisanal affair. 'Shoemakers' or 'cordwainers' (cobblers being those who repair shoes) produce a range of footwear items, including shoes, boots, sandals, clogs and moccasins – from a vast array of materials.

When people started wearing shoes, there were only three main types: open sandals, covered sandals and clog-like footwear. The most basic foot protection, used since ancient times in the Mediterranean area, was the sandal, which consisted of a protective sole, attached to the foot with leather thongs. Similar footwear worn in the Far East was made from plaited grass or palm fronds. In climates that required a full foot covering, a single piece of untanned hide was laced with a thong, providing full protection for the foot, thus forming a complete covering. These were the main two types of footwear, produced all over the globe. The production of wooden shoes was mainly limited to medieval Europe however – made from a single piece of wood, roughly shaped to fit the foot.

A variant of this early European shoe was the clog, which were wooden soles to which a leather upper was attached. The sole and heel were generally made from one piece of maple or ash two inches thick, and a little longer and broader than the desired size of shoe. The outer side of the sole and heel was fashioned with a long chisel-edged implement, called the clogger's knife or stock; while a second implement, called the groover, made a groove around the side of the sole. With the use of a 'hollower', the inner sole's contours were adapted to the shape of the foot. In even colder climates, such designs were adapted with furs wrapped around the feet, and then sandals wrapped over them. The Romans used such footwear to great effect whilst fighting in Northern Europe, and the native Indians developed similar variants with their ubiquitous moccasin.

By the 1600s, leather shoes came in two main types. 'Turn shoes' consisted of one thin flexible sole, which was sewed to the upper while outside in and turned over when completed. This type was used for making slippers and similar shoes. The second type united the upper with an insole, which was subsequently attached to an out-sole with a raised heel. This was the main variety, and was used for most footwear, including standard shoes and riding boots.

Shoemaking became more commercialized in the mid-eighteenth century, as it expanded as a cottage industry. Large warehouses began to stock footwear made by many small manufacturers from the area. Until the nineteenth century, shoemaking was largely a traditional handicraft, but by the century's end, the process had been almost completely mechanized, with production occurring in large factories. Despite the obvious economic gains of mass-production, the factory system produced shoes without the individual differentiation that the traditional shoemaker was able to provide.

The first steps towards mechanisation were taken during the Napoleonic Wars by the English engineer, Marc Brunel. He developed machinery for the mass-production of boots for the soldiers of the British Army. In 1812 he devised a scheme for making nailed-boot-making machinery that automatically fastened soles to uppers by means of metallic pins or nails. With the support of the Duke of York, the shoes were manufactured, and, due to their strength, cheapness, and durability, were introduced for the use of the army. In the same year, the use of screws and staples was patented by Richard Woodman. However, when the war ended in 1815, manual labour became much cheaper again, and the demand for military equipment subsided. As a consequence, Brunel's system was no longer profitable and it soon ceased business.

Similar exigencies at the time of the Crimean War stimulated a renewed interest in methods of mechanization and mass-production, which proved longer lasting. A shoemaker in Leicester, Tomas Crick, patented the design for a riveting machine in 1853. He also introduced the use of steam-powered rolling-machines for hardening leather and cutting-machines, in the mid-1850s. Another important factor in shoemaking's mechanization, was the introduction of the sewing machine in 1846 – a development which revolutionised so many aspects of clothes, footwear and domestic production.

By the late 1850s, the industry was beginning to shift towards the modern factory, mainly in the US and areas of England. A shoe stitching machine was invented by the American Lyman Blake in 1856 and perfected by 1864.

Entering in to partnership with Gordon McKay, his device became known as the McKay stitching machine and was quickly adopted by manufacturers throughout New England. As bottlenecks opened up in the production line due to these innovations, more and more of the manufacturing stages, such as pegging and finishing, became automated. By the 1890s, the process of mechanisation was largely complete.

Traditional shoemakers still exist today, especially in poorer parts of the world, and do continue to create custom shoes. In more economically developed countries however, it is a dying craft. Despite this, the shoemaking profession makes a number of appearances in popular culture, such as in stories about shoemaker's elves (written by the Brothers Grimm in 1806), and the old proverb that 'the shoemaker's children go barefoot.' Chefs and cooks sometimes use the term 'shoemaker' as an insult to others who have prepared sub-standard food, possibly by overcooking, implying that the chef in question has made his or her food as tough as shoe leather or hard leather shoe soles. Similarly, reflecting the trade's humble beginnings, to 'cobble' can mean not only to make or mend shoes, but 'to put together clumsily; or, to bungle.'

As is evident from this short introduction, 'shoemaking' has a long and varied history, starting from a simple means of providing basic respite from the elements, to a fully mechanised and modern, global trade. It is able to provide a fascinating insight not only into fashion, but society, culture and climate more generally. We hope the reader enjoys this book.

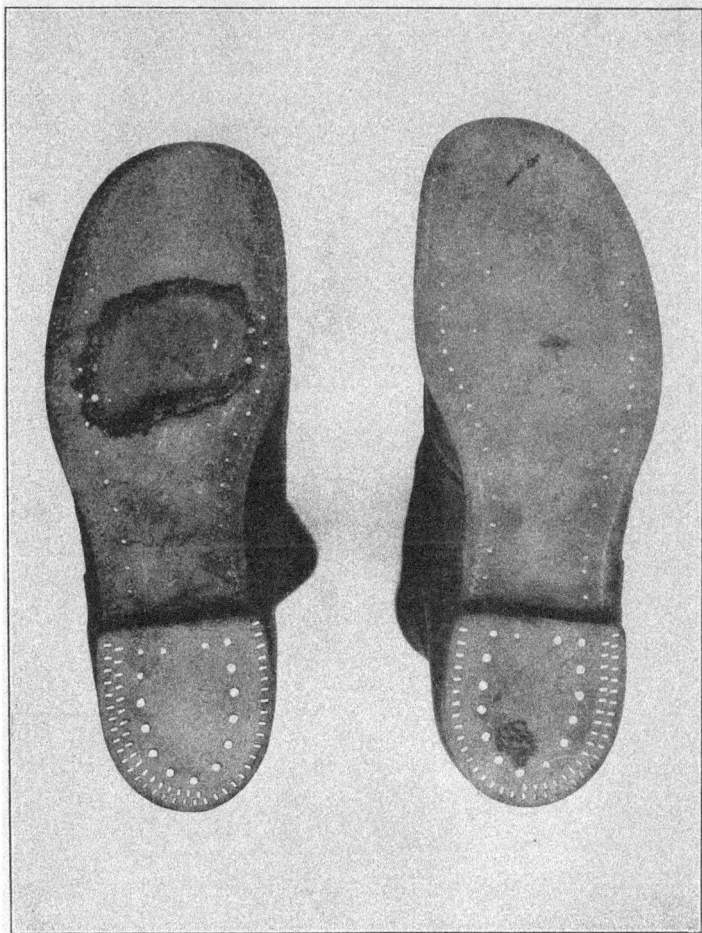

FIG. 25.—*Typical appearance of soles after completion of actual service test*

EFFECTS OF GLUCOSE AND SALTS ON THE WEARING QUALITY OF SOLE LEATHER

By P. L. Wormeley, R. C. Bowker, R. W. Hart, L. M. Whitmore and, J. B. Churchill

CONTENTS

I. INTRODUCTION

The question as to what constitutes the best wearing sole leather is one which vitally affects nearly every individual in the country. The extent of production of leather for shoe soles in this country alone is clearly indicated by the fact that in normal times the hides of 9 000 000 cattle are tanned and finished annually for this purpose. From these 9 000 000 hides which are equivalent to 18 000 000 sides[1] or bends, there are produced the enormous total of 504 000 000 pairs of men's, women's and children's soles, which represent a per capita production of approximately five pairs.

---

[1] Census of Manufacturers, 1914—The Leather Industry, Bureau of the Census, Department of Commerce.

The demand for sound and high quality sole leather was greatly increased by the participation of this country in the war. It was necessary that our soldiers be shod only with the best obtainable leather and to accomplish this, efforts were made to compile a suitable specification under which the leather could be purchased. It proved to be a difficult matter to present a specification satisfactory to every one concerned, because of honest differences of opinion which developed among tanners, chemists, manufacturers and other men in the trade, in regard to just what should be the proper composition of good quality sole leather. Several points of difference arose, chief among which was the effect of the presence of added glucose and salts on the wearing quality of the leather. This point was made a subject of investigation and this paper, the first of a series, deals with the methods and results of the tests conducted.

During the latter part of 1917, samples of 44 brands of commercial oak sole leather were analyzed at the Bureau of Standards for the Council of National Defense, which then controlled the purchases of shoes for the War Department. The analyses showed a variation in the water-soluble content of from 15 to 28 per cent, in glucose from 1 to 11 per cent, and in salts from practically none to 5 per cent. Two main classes of leather were represented, viz, one that contained little or no added glucose and salts and one that contained large quantities of added glucose and salts. The advocates of leather with a high percentage of glucose and salts claimed that the process allowed more tanning material to be added without producing a harsh, cracky leather, that it plumped the hide, thereby causing increased thickness and hence wear, thus actually conserving the hide supply, and produced a well filled, firm leather from which the glucose and salts would not wash out faster than the leather itself wore away. It was further pointed out that the experience of shoe manufacturers showed that the addition of at least a certain amount of glucose and salts produced a leather of uniform color, a better "feel," and at the same time stiffened the flanky and loose fibered portions of the hide, thus allowing soles to be cut from parts of a hide which otherwise would be too soft and spongy for satisfactory use. Those who opposed the use of glucose and salts in sole leather claimed that it served no useful purpose, did not add to the life of the leather, and would readily wash out, leaving a sole less resistant to moisture, thereby hastening decomposition and injuring the health of the wearer. They further pointed out that this process was practised by many tanners for the purpose of weighting or adulterating the leather, which was accomplished by giving the hides a quick tannage and adding the glucose and salts to obtain the same weight at less cost than would be obtained by a longer tanning and the accompanying addition of tanning material.

It is proper to state at this point, by way of explanation, that a limited amount of glucose, which is the principal material used for filling sole leather, is present in many of the tanning materials used and that as much as 2 per cent[2] may be found in the finished leather from this source alone. Any amounts greatly in excess of this percentage indicate that some has been added in the finishing of the leather.

In view of the many conflicting opinions and lack of reliable information on the subject, the Bureau, with the cooperation of the War Department and the National Association of Tanners, conducted this investigation.

II. PURPOSE OF INVESTIGATION

1. PREVIOUS WORK

It is believed that previous to the time this investigation was started no experiments had been conducted on a large scale which gave results indicative of the effects of glucose and salts on the wear of sole leather. Many wearing tests have been made where only a few pairs of soles were used and the value of the results obtained was uncertain, because the location on the hide from which the soles were cut was not considered. The length of wear of any particular sole for any one type of leather depends largely upon its location on the hide. Many purely chemical investigations have been made on various classes of leather to show that they contained excessive amounts of weighting material, but no tests were made to show the effects of this same material on the wearing quality.

2. REASONS FOR FURTHER INVESTIGATION

Where the production of a commodity is so great, as in the case of sole leather, and when an article is of such universal use, it is but natural that considerable thought and effort should be given to the producing of a high quality material. Scientific investigations are largely responsible for the development of high grade products and thus an investigation of this character is of value

---

[2] Leather Industries Laboratory Book, by H. R. Proctor, 1908.

in so far as the results aid the manufacturer in the betterment of his product. The reasons for further investigation are that there are differences of opinion regarding the question involved among responsible men in the trade, that there is no information available which gives reliable data on the subject, and that it is desirable to promote in the leather trade a more active interest regarding the composition of sole leather.

3. DEFINITE OBJECT OF THE INVESTIGATION

The primary object of this particular investigation was to determine the effects of glucose and salts on the wearing quality of sole leather. In other words, an attempt was to be made by actual wear tests to see whether a leather containing a large amount of added glucose and salts would wear a longer or shorter time than a leather well tanned and filled with tanning materials. The important test to determine the durability of sole leather is the service test, but chemical investigations were also made to show the composition of the leathers tested.

At the same time other items of interest were to be studied, such as the variation in wear due to the location on the hide, an observation as to whether the water-soluble materials wash out of the leather, the water-resisting qualities of the various leathers, the effects of any other chemical constituents on the wearing quality, the effects of wear on the chemical composition of the leather, and the relative cost of the two types of leather tested.

III. METHODS

The investigation consisted of three distinct divisions of work, as follows: The selection of the leather, the field tests, and the laboratory tests. The selection of the leather consisted in securing brands which had the proper composition for the test. The field tests involved the actual wearing of the soles to determine the comparative wear of the different leathers. The laboratory tests conducted were wearing tests on a machine designed to give an indication of the relative durability of sole leather, and included also the determination of specific gravities, water absorption qualities, and chemical analyses of the original leather and the worn soles.

1. SELECTION OF THE LEATHER

The selection of suitable leather for the tests was a matter of considerable difficulty because there were so many tannages and brands available. In order to limit the variable due to the tannage in so far as possible, it was decided to limit the investigation to vegetable-tanned leather made in a manner to produce the nearest approach to purely oak-tanned leather. The two classes to be secured were one with no added glucose or salts and one with a large amount of these materials added. It was assumed that the former leather would have a low water-soluble content, while that of the latter would be much higher in comparison. Upon this basis samples of eight brands of commercial sole leather were obtained by first laying aside 15 bends from a carload lot of the leather as soon as it reached the warehouse so that the same leather would be available later should it be desired for test purposes. Samples from a few of the bends in each lot were then analyzed in order to make a study of the water solubles, and especially the glucose content. This latter constituent was found to vary from 1½ to 10 per cent, while only two of the leathers indicated that a large amount had been added. The surprising fact, however, was that the percentage of water solubles was approximately the same for all brands, which showed that those leathers not filled with soluble non-tannins contained sufficient soluble tanning materials to nearly equalize the water-soluble content. Four of these eight leathers were chosen for the test, two which indicated that considerable glucose had been added and two which showed that very little was present.

(*a*) DESCRIPTION OF THE LEATHER SELECTED.—Each brand of leather was made by a different manufacturer, and the processes used as obtained from the manufacturers are described hereafter and are representative of the methods now used in this country to produce a high grade leather.

Sample A was a leather which contained a high percentage of added glucose and no added salts. The hides used were Chicago Packer and South American Frigorifico steers. The preparation of the hides in the beam house was conducted in the usual manner with no special treatment. The hides were given a preliminary tannage in the rockers and then handled either four or five times in the layaways, according to the weight of the hides. The tanning materials used were chestnut oak bark, liquid chestnut wood extract, and a small quantity of dry quebracho. The manufacturers considered that they used a larger proportion of oak bark than is customary among tanners. After the tanning process had been completed the leather was scoured by washing and then running through a Quirin press, after which glucose was added by drumming, followed by the addition of oil by the same method.

The mechanical finishing processes consisted of setting out on a drum setting machine and then rolling twice. The resulting leather had the characteristic light oak color.

Sample B contained added glucose but in smaller proportion than was found in sample A. Salts also were added. The Bureau was unsuccessful in obtaining from the manufacturers any information regarding the treatment of this leather.

Sample C was a leather containing a small amount of added glucose and salts for the purpose only of producing uniform color. The hides were La Plata South American steer, January kill. The usual beam-house methods were employed in the preparation for tanning. The tanning was accomplished by first hanging in handlers, rocking, and then laying away. After the vat tannage had been completed to the belting stage the hides were further tanned by adding the maximum amount of extract possible in drums. The tanning agents were chestnut oak bark and chestnut wood extract. A Fitz-Henry machine was used for scouring, after which a small amount of glucose and salts was added in the oil mill previous to adding the oil in the same mill. After drying, the mechanical treatments given.consisted of first sammying the leather and then rolling. After partially drying, another rolling was given.

Sample D contained only a little glucose and salts. The hides were Armour heavy Texas steers, July kill. The usual beam-house methods were used with the exception of deliming, which was accomplished by the use of hen bate. No acids or chemical agents of any kind were used in the preparation of the hides for tanning, which process was first to suspend the hides in the handlers, after which they were given a full six-months' treatment by the old-fashioned lay-away process. The tanning materials used were chestnut oak bark with a small percentage of chestnut wood extract. After tanning, the excess of soluble and insoluble materials was removed by scouring on a Fitz-Henry machine, followed by bleaching. After oiling in a wheel the leather was dried. The mechanical finishing consisted of sammying the leather and sponging it with clear water and cod oil, after which a heavy rolling was given. The leather was then resammied until the next day and given a final rolling. A thorough inspection was then made and any parts not well rolled were given additional treatment. By using the extreme pressure of the rolls the fibers were well packed down so that a firm water-resisting leather was produced.

121192°—19——2

(*b*). PREPARATION OF SAMPLES.—The procedure outlined for conducting the investigation not only called for actual service tests of the leather but also required several laboratory examinations, the results of which were to be correlated as far as possible with the field tests. Considerable attention was given to the proper cutting of the bends of leather so that satisfactory samples could be obtained for all tests and comparisons. Fig. 1 shows the manner in which each bend was divided into blocks, from each of which samples were obtained for the various tests. In all cases· the bends were marked off into strips 5 inches wide, beginning at the intersection of the shoulder end and backbone edge. The strips were then divided into blocks 15 inches long. These blocks were numbered consecutively across the bend from the

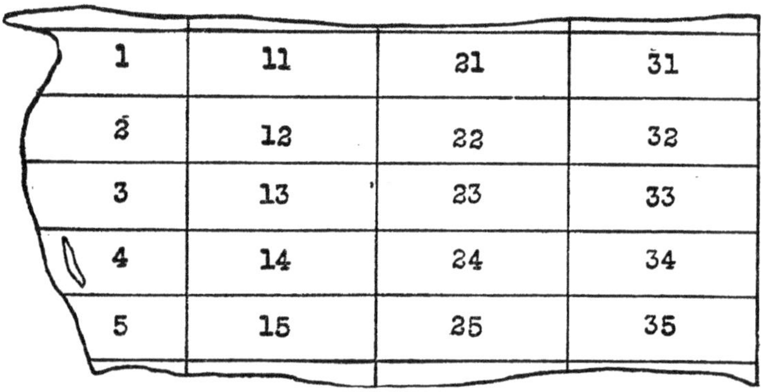

1	11	21	31
2	12	22	32
3	13	23	33
4	14	24	34
5	15	25	35

FIG. 1.—*Chart showing the method of dividing a bend into blocks*

backbone edge to the belly edge, beginning with No. 1 at the butt end near the tail, and in the direction from butt to shoulder were numbered in steps of 10. In this manner each block received a code number which definitely fixed its location on the bend. The bends of leather for each type were numbered from 1 to 15, inclusive, and all the blocks cut from one type of leather were stamped with a code letter. Thus a block bearing the symbol 5–B–23 would be readily recognized as being cut from bend No. 5 of type B in location 23. Fig. 2 represents the manner in which each individual block was divided.

Samples for Field Wearing Tests.—Each bend of leather was cut into blocks as described previously. From each of the blocks a sole (Fig. 2, *A*) was cut large enough to cover a range of shoe sizes up to and including size 11, care being taken to see that each sole was stamped with the same identification symbol appearing on the block from which it was cut. All soles were cut in the same direction, with the toe of the die pointing toward the shoulder end of the bend. An even number of left and right soles were cut from each type, so that all soles of any particular leather would not be worn on the same foot. After being matched in pairs the soles of each pair were evened to the same iron or thickness and were then ready for attaching to shoes for testing. Little skiving of any sole was required.

Samples for Machine Wearing Tests.—Each block was of sufficient size so that the part remaining beyond the toe of the die was large enough for a test specimen to be used on the laboratory wearing test machine. ˙ This sample is represented by Fig. 2, *B*.

Samples for Chemical Analyses.—The scrap around the sole represented by Fig. 2, *C*, was collected after each sole had been cut and placed in a container marked with the symbol on its

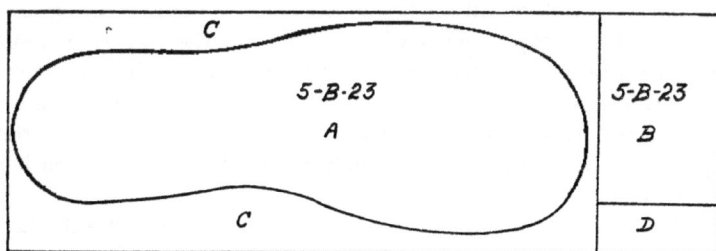

FIG. 2.—*Chart showing the division of a block into samples for field and laboratory tests*

respective block. A composite sample from all the blocks of the same location for each type of leather was used for chemical analyses.

Samples for Specific Gravity Tests.—Samples for the determination of specific gravity were obtained from that part of the block adjacent to the sole and wearing machine samples designated by Fig. 2, *D*.

Samples for Water Absorption Tests.—A few samples for this test were obtained from each location on the bend where the blocks were either not large enough to furnish a sole or were damaged by brands or cuts to such an extent that the sole was valueless for testing.

2. FIELD TESTS

(*a*) WEARING TESTS.—The most practical and satisfactory method of determining the durability of sole leather is by an actual test in service. To select a number of soles at random without knowledge of from what parts of the hides they were obtained and then place them in service indiscriminately among several individuals, the nature of whose work is radically different, introduces so many variables that the results of the tests will be of questionable value. With this in mind, the character of these tests was so planned that all possible variables were eliminated. The type of tannage was first considered and it was decided to use leather tanned with oak bark and chestnut wood extract. The fibrous structure varies greatly throughout any hide, those parts near the back and over the kidneys being close fibered and producing the firmest leather while the belly and shoulder parts are made up from looser fibers and thus produce leather which is sometimes soft and spongy, less resistant to moisture, less firm, and hence less durable. This difference in texture renders the location from which a sole is cut a most important and deciding factor in its durability. Other variable factors affecting the results of any experiments of this kind are the weight of individuals, physical imperfections, or personal peculiarities in walking, which cause more pressure to be placed on one foot than on the other, the nature of the work performed, the nature of the ground on which the individual habitually walks, and the weather conditions.

Plan of Comparison.—A system of making the tests was decided upon which was designed to eliminate or compensate for many of these varying conditions. The soles were so matched in pairs that each individual would make a complete test. This was accomplished by placing on one shoe of each pair a certain type of leather and on the other shoe the particular leather with which the comparison was desired. The method of compensating for the difference in wear due to location on the hide was to match the pairs so that the two soles of any pair were cut from the same portion of their respective bends. Thus a sole cut from location 23 on a bend of types A or B leather would be tested against a sole cut from location 23 on a bend of types C or D. Since there were 15 bends of each brand of leather there were 15 soles cut from the same location, 8 of which were right soles and 7 left soles. Half of the soles from each of samples A and B, which represented the leather filled with glucose and salts, were tested against sample C, while the other half were tested against sample D. If all the soles had been suitable for testing purposes there would have been 225 soles to each brand. This number was reduced con-

FIG. 4.—*Laboratory wearing machine*

siderably by unsatisfactory soles, but it was sufficiently great and the soles were matched in such a manner that the differences in wear due to individual characteristics were equalized in the average results. .

The men selected to make the test were infantry soldiers at Camp Meade, Md. The advantages of testing the soles on soldiers were that all men were as nearly as possible of the same physical standard, their work on duty was the same in each case, and all tests were made under the same ground and weather conditions.

System of Inspection.—A system of inspection was introduced whereby a representative of the Bureau visited the camp one day of each week to inspect the soles while the test was in progress. This inspection was readily made because of the fact that on the day designated all the men in each barracks who were wearing test shoes would assemble them in some convenient location in the building. This weekly examination gave an opportunity of keeping in touch with the test and of observing the performance and wear of the various leathers.

Records.—A careful record was made of the soles on each pair of shoes and a number was assigned to them which was punched on the tongue of the shoes. A weekly report card was designed and furnished each man wearing test shoes. On the face of the card appeared the date, number assigned to the pair of soles, test number, report number, and spaces provided for the wearer to fill in stating the days worn, the kind of work done, the weather conditions, and remarks. Fig. 3 represents the form used. The card for the week was placed in one of the shoes on inspection day and was collected by the Bureau representative, who left a similar form to be used the following week. When either sole was worn through the date was noted on the card and the number of days worn was then easily calculated.

3. LABORATORY TESTS

(*a*) MACHINE WEARING TESTS.—The field wear tests were supplemented by laboratory machine wearing tests to see whether this machine could be used to indicate the behavior of sole leather in actual service.

Description of Wearing Machine (Fig. 4).—A wheel of 15 inches diameter carries on its face twelve test pieces. The wheel revolves at the rate of 30 revolutions per minute about a horizontal axis with its bearings in two parallel metal bars which are pivoted at one end, the other end being free. The wheel carrying the weight of the bars (and any additional weight that may be suspended from their free end) rests on a horizontal disk of 16 inches diameter, the point of contact being 5 ½ inches from the axis of the disk. This disk has a surface of carborundum and rotates.about a vertical axis on which is a brake wheel provided with a brake strap, by means of which any desired resistance to rotation may be secured by the application of dead weight. The wheel is driven by a chain, and in turn drives the horizontal disk with which the test pieces are in contact. The apparatus is designed with the view of subjecting the test piece to (1) a driving (shearing) action under pressure and (2) a slight abrasive action resulting from the

DEPARTMENT OF COMMERCE
BUREAU OF STANDARDS
Form 388

WEEKLY REPORT OF WEARING TESTS ON SHOES

Pair No. .... Test No. .... Report No. ....

For week ending ...., 1918.

CHECK (√) DAYS WORN	DUTY	WEATHER CONDITIONS
Sunday...............		
Monday		
Tuesday...............		
Wednesday...............		
Thursday...............		
Friday...............		
Saturday		
Remarks:		

1)—6403

FIG. 3.—*Record card for field wearing tests*

circular path of contact between the wheel and disk. The conditions of pressure and shear may be adjusted as desired.

A circular brush is shown resting on the carborundum disk. This brush in connection with a small exhauster tends to keep the surface of the wearing disk clean.

A test consists of 40 000 revolutions of the wheel, which corresponds with 40 000 steps, or approximately 40 miles of walking.

Method of Making Test.—Rectangular samples, about 4 inches long and 2 inches wide, of the leather to be tested are prepared, carefully weighed, and then attached to the face of the wheel by means of countersunk screws. The wheel was designed to allow space for twelve samples. Between each leather sample a specimen of rubber composition was inserted which acted in a manner to prevent the small interstices of the carborundum surface from being clogged with leather dust. Thus six samples, representing the leather used on three pairs of shoes, were tested at one time. The wearing test consists simply of allowing the test wheel to revolve for 40 000 revolutions, after which enough material has worn away to give an indication of the relative wear. The samples are then weighed again to determine the loss in weight. The loss in volume is then determined from the loss in weight and the specific gravity. The leather showing the least loss in volume should have the longest life in service.

(*b*) CHEMICAL ANALYSES.—A complete chemical analysis was made of the leather from each location on the bend (Fig. 1) for each brand. For example, material from location number 23 of each bend of the same kind of leather was used for a composite sample for the analysis of that block. A similar sample was obtained for all locations of each brand making a total of 80 complete analyses.

A similar analysis was made of the worn soles to determine the effects of the wear on the composition of the leather.

The chemical work was done cooperatively by the leather laboratory of the Bureau and the American Leather Research Laboratory. The official methods of the American Leather Chemists' Association were followed.

(*c*) DETERMINATION OF SPECIFIC GRAVITY.—A determination of specific gravity was required for use in calculating the results of the machine wearing tests and for the purpose of obtaining information as to the densities of the different brands of leather tested.

Apparatus and Method Used.—The instrument used for determining the specific gravity was a direct reading gravitometer (Fig. 5). Small samples of leather were coated with cellulose nitrate to make them waterproof. A sample was then attached to the needle point of the apparatus and the weight beam moved to bring the index opposite the upper limit of the graduated arc. The glass vessel, filled with water, was then raised until the sample was completely immersed. The pointer then moved along the scale and indicated the specific gravity. The determinations were accurate to the second decimal place.

(*d*) WATER ABSORPTION TESTS.—The resistance of leather to the penetration of moisture depends on several conditions. The compactness of the fibers, amount of soluble materials present, tannage, and amount of grease present are all factors which have an influence on this property.

Method of Making Tests.—Samples were prepared 1¾ inches square and carefully weighed. They were then immersed in distilled water in a suitable receptacle for 30 minutes, after which time they were wiped dry and weighed again. The samples were then placed in a fresh supply of distilled water and allowed to remain for 24 hours, after which another weighing was made. In order to arrive at the true percentage of moisture absorbed a

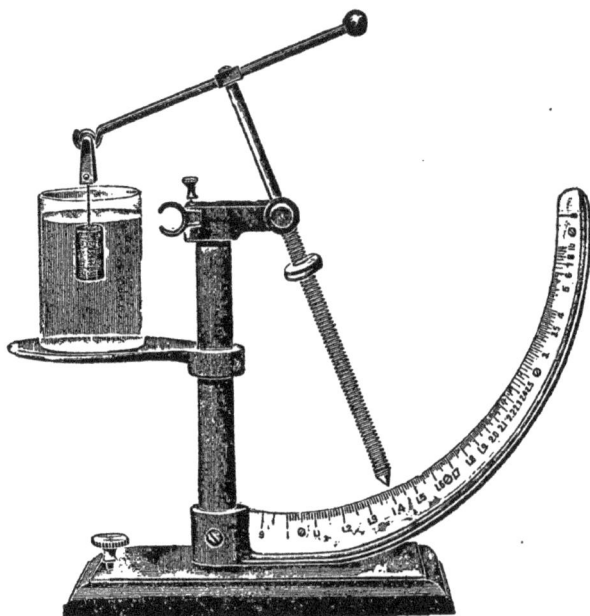

FIG. 5.—*Direct reading gravitometer*

correction was made for the soluble materials that soaked from the leather, by evaporating the solutions in which the samples were soaked and adding the percentage thus found to have leached out, to the apparent water absorption.

IV. DATA AND RESULTS OBTAINED

1. FIELD TESTS

The results of the actual service tests are shown graphically in Fig. 6, where the days wear per iron or unit of thickness is plotted against the location on the hide from which the soles were cut. The values represent the average days wear per iron for the total number of soles tested from any particular location. The average number of soles from which results were obtained was seven. The values for days wear per iron were obtained by dividing the sum of the number of days wear obtained from all the soles in a similar location of each brand by the sum of the irons for the same soles. The iron is a unit of measure used in the trade to designate the thickness or gage of leather and is equal to approx-

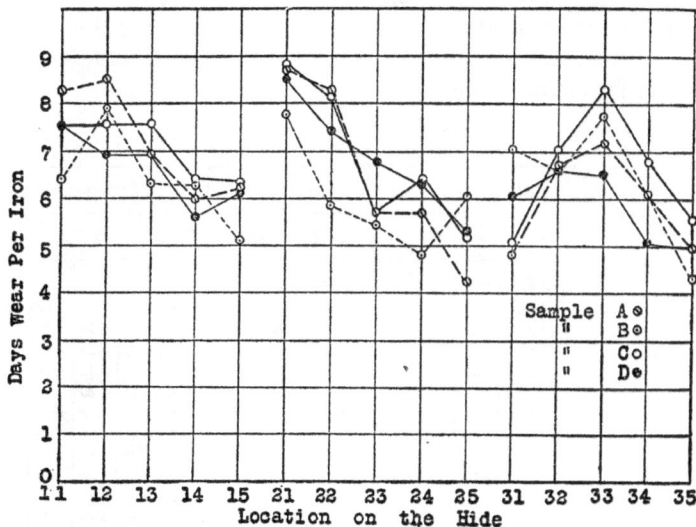

Fɪɢ. 6.—*The wear expressed in days per iron for all soles from each location on the hide for the four brands*

imately ½mm in the metric scale, and equal to 1/48 inch in the English scale.

Fig. 6 shows three groups of curves which represent the soles from locations 11 to 15, 21 to 25, and 31 to 35, inclusive.

(a) VARIATION IN WEAR OF DIFFERENT LEATHERS.—From Fig. 6 it will be observed that the wear of one particular brand of leather was not consistently greater than that of any other brand throughout all the locations on the hide. Thus it can be stated that the wear of all the brands tested was nearly the same. Table 1 shows the results obtained for the days wear per iron, days wear per sole, and average iron per sole for each location of each sample. Samples C and D were the leathers with little or no added glucose and salts, while A and B were leathers with one or both of these materials added. The difference in days wear per sole is nearly accounted for by the difference in the average iron per sole. The difference in the days wear per iron of leathers C and D as compared with A and B is only 3 per cent, which is a small difference inasmuch as the nature of the test might well cause an experimental error of this or even a greater amount. The results used in plotting the graphs were only those which were definitely known to be correct. In 103 cases out of 213 pairs from which data were secured, leathers C or D wore the longer time, in 74 cases leathers A or B wore the longer time, and in 36 instances the length of wear was the same for both types of leather.

TABLE 1.—Wear Data

Location	Days wear per sole				Average iron per sole				Days wear per iron			
	A	B	C	D	A	B	C	D	A	B	C	D
11......................	85.5	62.8	76.1.	74.3	10.30	9.80	10.14	9.86	3.31	6.41	7.52	7.55
12......................	94.0	81.2	79.0	74.3	11.00	10.25	10.40	10.66	8.54	7.92	7.59	6.96
13......................	77.3	70.8	83.7	78.8	11.02	11.16	11.00	11.33	6.95	6.34	7.60	6.96
14......................	67.3	68.0	70.5	63.0	11.02	10.80	11.00	11.20	6.00	6.29	6.42	5.62
15......................	61.5	47.0	59.8	59.5	9.85	9.20	9.40	9.66	6.25	5.12	6.36	6.15
21......................	81.6	74.8	83.3	81.0	9.33	9.60	9.50	9.50	8.75	7.80	8.76	8.52
22......................	84.8	60.3	85.2	74.7	10.20	10.25	10 40	10.00	8.32	5.88	8.18	7.47
23......................	61.8	59.0	59.2	72 8	10.62	10.57	10.33	10.70	5.72	5 58	5 73	6.80
24......................	55.9	46.5	64.0	59.3	9.80	9.62	11.43	9.40	5.72	4.83	6.41	6.31
25......................	38.8	56.8	49.0	48.5	9.12	9.33	9.37	9.12	4.26	6.08	5.23	5.32
31......................	41.3	65.6	44.5	54.7	8.50	9.26	8.83	9.00	4.82	7.08	5.04	6.09
32......................	60.6	60.0	63.8	58.6	9.00	9.00	9.11	8.88	6.74	6.67	7.02	6.61
33......................	68.1	72.9	78.2	61.9	9.50	9.40	9.40	9.50	7.18	7.76	8.32	6.52
34......................	55.9	56.6	61.7	48.8	9.14	9.26	9.12	9.33	6.10	6.09	6.78	5.09
35......................	40.3	36.4	45.3	39.8	8.11	8.43	8.22	8.28	4.97	4.32	5.53	4.98
Average.................	64.8	61.2	68.9	63.2	9.77	9.73	9.83	9.75	6.57	6.28	6.83	6.46

(*b*) VARIATION IN WEAR AT DIFFERENT LOCATIONS ON THE HIDE.—Fig. 6 shows very clearly that there is a variation in the wear due to the location on the hide from which the sole was cut. The best wearing parts of the hide are represented by the shaded portion on Fig. 7 which includes blocks 11, 12, 13, 21, 22, 23, 32, and 33. This area includes the firm solid part of the hide near the back and over the kidneys, extending the entire length of the bend and across the hide to about 15 inches from the backbone, excluding, however, a small portion of the hide on the shoulder end near the backbone. The soles from this location (block 31) showed a low wearing quality in nearly every case. The curves in Fig. 6

show a tendency for the wear to be best for those soles near the backbone edge, except at the shoulder end, and for the wear to be poorer for those soles near the belly edge where the fibres are less compact. From these results it is apparent that the location of the sole in the hide is far the most influential factor governing its wearing quality.

2. LABORATORY TESTS

(*a*) MACHINE WEARING TESTS.—The differences in wear in service between the four brands of leather tested were so small that the machine tests were not expected to show any very great variation in wear between the leathers. The machine tests did,

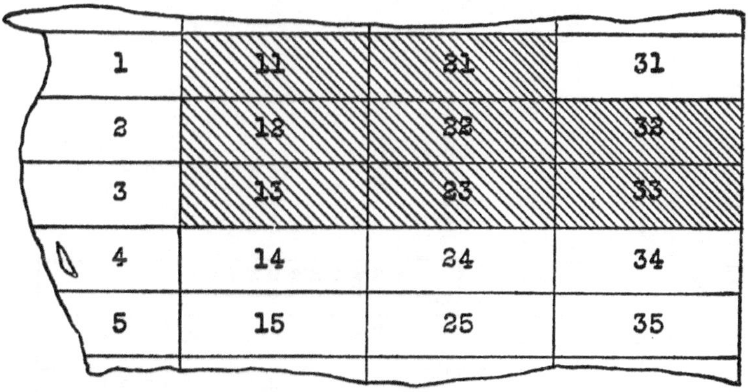

1	11	21	31
2	12	22	32
3	13	23	33
4	14	24	34
5	15	25	35

FIG. 7.—*Shaded portion shows location of best wearing leather on the hide*

however, give a general indication of the relative wear of the soles from the different locations on the hide.

(*b*) CHEMICAL ANALYSES OF ORIGINAL LEATHER.—A complete chemical analysis was made of each brand of leather for each location on the hide. The results are presented numerically in Table 2 and graphically in Figs. 8 to 17, inclusive. In the case of the graphs the values for each chemical constituent were plotted against the location on the hide. All chemical results were changed to a 12 per cent moisture basis for comparison.

Water-Soluble Material.—The water-soluble material consists of tannins and nontannins, the latter including any added glucose or salts. Glucose and salts form the chief nontanning materials used, and are more readily soluble than the tannins.

Technologic Papers of the Bureau of Standards

TABLE 2.—Data Calculated to 12 Per Cent Moisture

SAMPLE A

Block	Water-soluble		Glucose		Ratio of tannins to nontannins		Total ash		Insoluble ash		Epsom salts		Petroleum ether extracts		Hide substance		Combined tannin		Degree of tannage	
	New	Worn	New	Worn	New	Worn	New	Worn	New	Worn	New	Worn	New	Worn	New	Worn	New	Worn	New	Worn
1	25.6		11.5		0.65		0.36		0.14		0.12		2.3		35.2		24.8		70	
2	22.6		8.3		.79		.31		.15		.12		1.7		38.5		24.1		63	
3	22.2		8.2		.72		.33		.15		.18		1.7		37.9		26.1		69	
4	24.4		9.8		.68		.36		.18		.14		2.3		36.8		24.4		66	
5	25.5		12.5		.62		.38		.19		.12		3.0		36.0		23.4		65	
11	21.7	16.8	7.2	2.8	.78	0.93	.29	2.44	.13	1.41	.12	0.66	1.7	3.6	38.4	40.8	26.1	25.4	68	62
12	20.4	12.6	7.6	2.0	1.16	.64	.25	2.31	.13	1.44	.10	.42	1.3	3.0	40.6	43.3	25.8	27.3	64	63
13	20.3	14.6	7.5	1.0	.79	.93	.25	2.63	.12	1.69	.11	.66	1.3	3.7	39.5	41.0	26.8	27.0	68	66
14	21.3	13.6	7.5	1.3	.80	.78	.28	3.06	.14	1.90	.11	.29	1.5	3.5	38.6	39.4	26.5	29.1	69	74
15	23.4	13.6	9.6	0.5	.69	1.04	.32	2.83	.18	1.53	.11	.69	2.1	3.7	36.8	41.9	25.6	27.3	70	65
21	22.5	11.8	9.0	1.6	.73	.69	.29	2.39	.15	1.33	.11	.51	1.7	3.6	38.1	38.1	25.6	32.7	67	86
22	21.2	15.2	7.0	2.5	.78	1.05	.28	2.79	.15	1.58	.11	.97	1.1	2.7	39.0	39.0	26.6	29.5	68	76
23	21.1	13.6	8.3	1.3	.90	.74	.27	3.12	.14	1.80	.12	.60	1.4	4.0	38.1	36.4	27.3	31.6	72	87
24	22.4	15.7	8.5	1.0	.75	.80	.29	3.68	.14	2.33	.12	.94	1.6	3.7	38.0	36.4	25.9	29.9	68	82
25	22.8	13.5	10.3	1.2	.75	.93	.32	4.56	.15	2.94	.13	.62	2.1	4.3	36.6	38.4	26.4	28.1	72	73
31	22.8	13.9	7.6	1.6	.79	.92	.30	3.56	.14	2.31	.12	.48	1.5	4.5	37.3	37.6	26.3	29.4	71	78
32	21.1	15.7	7.0	2.0	.83	.83	.29	3.07	.13	1.76	.12	.75	1.5	4.1	37.9	38.0	27.4	28.4	72	75
33	21.6	12.9	9.2	1.3	.79	.88	.28	3.70	.12	2.29	.13	.65	1.7	3.9	38.7	38.1	25.9	30.2	67	79
34	21.8	15.8	8.2	0.0	.77	.97	.29	3.22	.14	2.21	.13	.42	1.6	4.9	38.7	37.0	25.8	28.1	67	76
35	25.5	13.0	11.4	1.2	.68	.85	.37	4.82	.18	3.16	.14	.48	2.7	5.9	35.6	36.8	24.1	28.5	68	77
Average a	22.5		8.8	1.4	.77		.31		.15		.12		1.8		37.8		25.7		68	
Average b	22.0	14.1	8.4		.80	.86	.29	3.21	.14	1.98	.12	.61	1.6	3.9	38.1	38.8	26.1	28.8	69	74

SAMPLE B

	C1	C2	C3	C4	C5	C6	C7	C8	C9	C10	C11	C12	C13	C14	C15	C16	C17	C18	C19	C20
1	23.2		4.3		0.99		1.20		0.35		2.03		4.1		34.2		26.2		77	
2	22.0		4.4		1.05		1.27		.38		1.87		3.5		33.8		28.4		84	
3	22.1		3.9		1.05		1.14		.34		1.96		2.8		34.0		28.8		85	
4	24.0		5.0		.96		1.36		.44		2.02		3.8		32.5		27.3		84	
5	25.1		4.4		.89		1.63		.48		2.73		4.8		30.8		26.8		87	
11	21.8	22.6	3.3	0.6	1.23	1.34	.95	3.19	.31	1.88	1.60	0.96	3.7	4.4	37.1	34.0	25.1	25.1	68	74
12	20.8	20.0	2.7	1.3	1.18	1.07	.90	2.70	.26	1.39	1.46	.53	2.9	3.6	37.8	37.0	26.2	25.5	69	69
13	21.7	21.9	2.8	1.3	1.19	1.18	.85	2.79	.25	1.38	1.33	.93	2.6	4.0	38.4	36.5	25.1	24.2	65	66
14	22.2	18.3	2.6	1.7	1.21	1.12	.90	3.26	.27	1.87	1.56	.60	2.1	4.0	38.0	36.9	25.4	26.3	67	71
15	23.7	22.4	4.9	0.5	1.11	1.33	1.14	2.75	.32	1.76	1.97	1.07	2.8	3.0	34.3	35.0	26.9	25.8	78	74
21	22.1	19.5	3.5	2.5	1.15	1.04	.97	2.15	.30	1.00	1.82	.95	2.5	3.6	36.5	37.1	26.6	26.3	73	71
22	21.9	21.7	3.7	1.3	1.21	1.05	.87	2.35	.26	1.09	1.38	.48	2.4	2.4	36.9	37.3	26.5	25.5	72	68
23	22.2	21.4	3.5	1.3	1.07	1.07	.87	7.51	.22	1.23	1.41	.87	2.2	4.4	37.5	34.3	25.9	26.1	72	76
24	22.0	19.2	3.4	0.0	1.30	1.17	.91	2.78	.27	1.43	1.53	.61	2.2	5.4	36.8	35.4	26.7	26.6	73	75
25	22.7	18.6	4.4	1.2	1.14	1.17	1.19	3.64	.29	1.96	2.15	.69	2.6	5.1	33.8	33.7	28.6	27.7	85	82
31	22.0	18.6	4.0	.6	1.04	1.10	1.09	2.69	.32	1.99	1.94	.68	2.7	3.8	35.9	36.0	27.1	27.6	75	77
32	20.8	18.5	3.5	1.7	1.25	1.05	.94	3.61	.28	2.07	1.64	.78	2.0	4.8	36.3	33.9	28.6	28.0	79	83
33	22.1	18.6	3.7	.9	1.17	1.05	.97	3.12	.29	1.87	1.66	.34	2.1	3.9	35.9	35.8	27.6	27.8	77	76
34	21.3	17.0	3.6	1.7	1.12	1.09	1.02	3.01	.30	1.55	1.77	.74	2.0	3.0	35.7	35.9	28.7	29.9	78	83
35	24.1	17.5	5.5	0.0	1.01	1.20	1.33	4.82	.38	2.23	2.30	.10	2.7	5.1	33.0	33.9	27.8	29.3	84	86
Average a	22.4		3.8		1.12		1.07		.31		1.80		2.8		35.5		27.0		76	
Average b	22.1	19.7	3.7	1.1	1.15	1.13	.99	3.02	.29	1.65	1.70	.69	2.5	4.0	36.2	35.5	26.8	26.8	74	75

a Analysis of all blocks.

b Analysis of all blocks on which tests were secured.

TABLE 2.—Continued.

SAMPLE C

Block	Water-soluble		Glucose		Ratio of tannins to nontannins		Total ash		Insoluble ash		Epsom salts		Petroleum ether extracts		Hide substance		Combined tannin		Degree of tannage	
	New	Worn	New	Worn	New	Worn	New	Worn	New	Worn	New	Worn	New	Worn	New	Worn	New	Worn	New	Worn
1	22.4		1.6		1.52		1.45		0.22		1.42		3.4		32.6		29.4		90	
2	21.8		1.3		1.63		1.33		.22		1.14		3.6		33.6		28.8		86	
3	23.9		1.6		1.54		1.47		.23		1.45		3.6		31.5		28.8		91	
4	24.0		1.7		1.55		1.47		.22		1.38		3.7		31.7		28.4		90	
5	28.0		2.5		1.46		2.00		.25		1.48		3.9		27.7		28.2		104	
11	21.0	20.8	1.3	1.3	1.50	1.24	1.25	2.56	.21	1.67	.93	0.86	3.3	3.5	34.8	35.9	28.7	26.1	83	74
12	20.4	21.8	1.3	1.2	1.75	1.26	.91	2.75	.18	1.39	.77	.46	2.4	3.3	36.5	35.9	28.5	25.6	78	71
13	20.3	19.7	1.2	1.2	1.74	1.17	.86	2.33	.18	1.60	.73	.98	1.9	4.0	37.2	37.0	28.4	25.7	76	69
14	20.8	21.2	1.3	1.1	1.74	1.18	.93	2.73	.18	1.88	.80	1.10	1.9	3.9	35.0	34.9	30.1	26.1	'86	75
15	22.2	18.5	2.0	1.5	1.62	1.12	1.26	3.10	.21	2.16	1.20	1.20	2.8	5.1	33.0	34.1	29.8	28.1	90	82
21	21.3	22.7	1.4	1.4	1.61	1.17	1.14	2.31	.18	1.39	1.12	.30	2.7	3.8	36.8	35.8	27.3	24.3	75	68
22	20.9	21.9	1.2	1.3	1.77	1.07	.98	2.50	.18	1.48	.94	1.31	1.7	5.5	36.1	35.7	29.1	23.4	81	66
23	21.3	22.5	1.0	1.3	1.73	1.07	.94	2.31	.17	1.66	.92	.48	1.9	4.1	36.4	36.4	28.2	24.3	78	69
24	21.2	21.3	1.3	1.5	1.75	1.14	.97	2.83	.17	1.84	1.00	.58	2.1	4.3	36.2	36.2	28.3	27.6	78	84
25	22.7	20.5	1.6	1.5	1.57	1.27	1.30	2.39	.18	1.63	1.15	.50	2.4	4.9	33.1	32.8	29.6	28.2	89	86
31	20.9	20.7	1.4	1.5	1.62	1.24	1.19	3.39	.18	2.38	1.23	.93	3.2	4.9	35.9	32.9	27.8	27.1	77	82
32	19.7	19.8	1.2	1.1	1.69	1.32	.96	3.27	.17	2.39	.87	.80	1.9	5.6	35.9	35.5	30.3	24.7	84	70
33	20.0	20.6	1.2	1.2	1.79	1.26	.84	2.94	.17	2.03	.76	.61	1.8	4.2	36.8	35.2	29.2	26.0	79	73
34	20.1	20.1	1.5	0.7	1.71	1.34	.95	2.91	.18	1.43	.92	.60	2.1	3.4	34.9	34.3	30.7	28.8	88	84
35	22.0	20.8	2.2	1.3	1.53	1.25	1.42	3.84	.22	2.85	1.23	.88	3.4	4.6	32.6	34.2	29.8	25.5	91	75
Average a	21.8		1.5		1.64		1.18		.20		1.07		2.7		34.4		29.0		84	
Average b	20.9	20.9	1.4	1.3	1.67	1.21	1.06	2.81	.18	1.85	.97	.77	2.4	4.3	35.4	34.8	29.1	26.1	82	75

SAMPLE D

Sample	1a	1b	2a	2b	3a	3b	4a	4b	5a	5b	6a	6b	7a	7b	8a	8b	9a	9b	10a	10b
1	76	…	26.9	…	35.3	…	2.5	…	0.54	…	0.12	…	0.39	…	1.18	…	2.6	…	23.2	…
2	74	…	26.9	…	36.1	…	2.3	…	.50	…	.12	…	.39	…	1.24	…	2.7	…	22.6	…
3	75	…	26.8	…	35.6	…	2.3	…	.48	…	.12	…	.39	…	1.25	…	3.0	…	23.2	…
4	84	…	27.5	…	32.8	…	3.0	…	.62	…	.12	…	.43	…	1.17	…	4.6	…	24.6	…
5	81	…	26.4	…	32.8	…	3.3	…	.60	…	.12	…	.47	…	1.17	…	3.8	…	25.4	…
11	72	72	26.8	26.2	37.2	36.2	2.2	3.9	.40	.59	.11	1.41	.30	2.47	1.36	1.12	3.0	1.9	20.8	20.3
12	73	60	27.6	22.6	37.6	37.8	1.9	4.4	.41	.32	.12	1.11	.32	2.29	1.47	1.31	2.3	2.6	21.6	22.1
13	72	63	27.0	24.4	37.7	38.7	1.6	3.4	.40	.55	.13	1.52	.35	2.54	1.43	1.17	2.3	1.8	22.3	20.0
14	73	62	26.9	23.6	37.0	37.9	1.7	4.2	.31	.77	.11	1.80	.36	2.38	1.36	1.40	3.8	1.4	24.1	20.5
15	74	74	26.1	25.8	35.5	37.0	2.7	3.4	.32	.68	.11	2.01	.31	3.30	1.27	1.27	3.2	1.4	22.4	21.9
21	70	65	26.3	23.8	37.5	36.5	1.7	3.3	.40	1.15	.11	1.36	.34	2.57	1.36	1.20	3.2	2.3	21.5	23.0
22	73	74	27.4	27.0	37.5	36.2	1.5	4.3	.41	.54	.12	1.82	.34	3.16	1.44	1.21	3.3	1.8	22.1	18.7
23	71	66	26.7	23.5	37.6	35.7	1.5	4.5	.42	.78	.12	1.31	.43	2.53	1.48	1.38	2.2	2.7	22.6	23.0
24	69	71	26.0	24.9	37.7	35.1	1.6	4.0	.50	.55	.11	1.69	.41	2.81	1.46	1.15	2.5	1.9	24.3	22.3
25	70	56	25.4	21.4	36.1	37.9	2.1	5.3	.36	.89	.11	2.34	.35	4.46	1.33	1.16	4.5	1.9	22.8	21.1
31	75	73	27.3	27.2	36.2	37.3	1.6	2.4	.44	.37	.12	1.37	.35	2.34	1.33	1.08	4.0	1.3	22.1	19.7
32	73	57	27.1	22.2	37.3	39.1	1.4	3.6	.44	1.24	.12	1.37	.36	2.56	1.56	1.32	3.1	1.5	22.1	21.7
33	72	73	27.0	26.5	37.3	36.1	1.5	2.6	.45	.85	.11	1.67	.37	2.86	1.56	1.05	1.8	1.4	22.3	21.1
34	79	59	28.4	22.9	35.9	39.1	1.3	3.9	.35	.87	.13	1.58	.40	2.85	1.48	1.33	3.7	2.0	24.4	20.5
35	76	75	26.6	26.0	34.9	34.6	2.0	5.9	.53	.60		2.65		4.09	1.30	1.28	4.2	1.6		18.9
Average [a]	74	…	26.9	…	36.3	…	2.0	…	.44	…	.12	…	.37	…	1.36	…	3.2	…	22.8	…
Average [b]	73	66	26.8	24.5	36.8	36.8	1.7	3.9	.41	.72	.11	1.67	.35	2.88	1.41	1.23	3.1	1.8	22.4	21.2

[a] Analysis of all blocks.

[b] Analysis of all blocks on which tests were secured.

The two classes of leather selected showed very little difference in the total water-soluble materials, the values varying from 21.75 to 22.8 per cent. Fig. 8 shows the distribution of the water solubles over the hide. It is highest in those portions of the hide that are loosely fibered and shows a tendency to be high on the

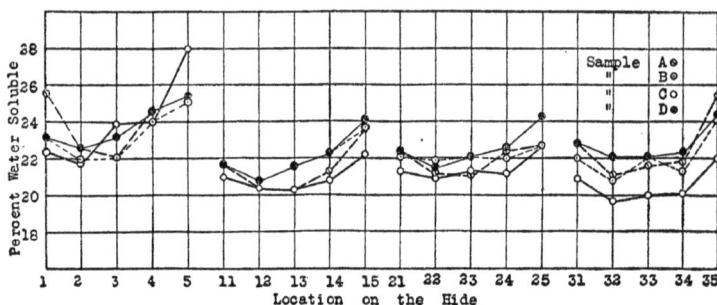

FIG. 8.—*Chart showing distribution of water soluble materials over the hide*

backbone edge, then drops slightly, after which it gradually increases until the maximum amount is reached in the portions along the belly edge. The fact that the water solubles are higher along the backbone edge than in the middle portions is probably accounted for by the fact that a greater opportunity was given for

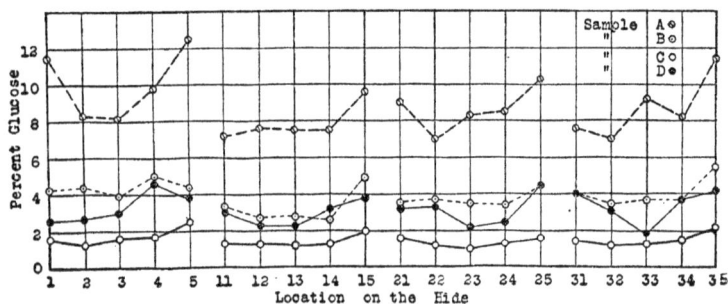

FIG. 9.—*Chart showing distribution of glucose over the hide*

absorption because the edge of the leather was exposed. This condition is clearly shown by the values for locations 1 to 5, which represent the extreme butt end of the bend. Locations 1 and 5 are corner pieces and hence have two edges exposed. The amount of surface exposed and the texture of the various parts of the leather influence the composition at any location on the hide.

Glucose.—As has been previously stated, glucose is used in varying quantities in finishing sole leather. Vegetable tanning materials may contain some sugar which is absorbed in the tanning processes, and thus it is probably impossible to find commercial leathers entirely free from this material. The small amounts found naturally in tanning materials are of great value in that it

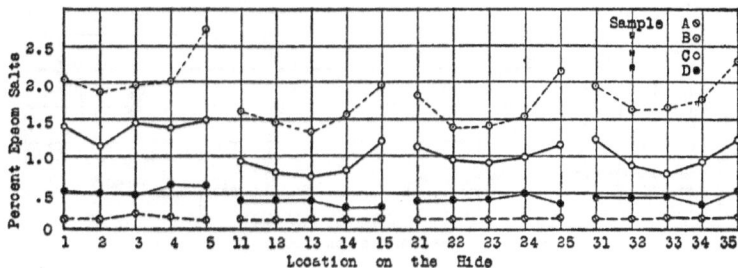

Fig. 10.—*Chart showing distribution of Epsom salts over the hide*

is from these sugars in part that the acid or plumping qualities of the liquors are derived. A very small quantity added artificially is said to give the leather a better finish and appearance and also adds solidity to the flanky and thin portions of the hide.

Fig. 9 shows the variation in glucose content between the different leathers, and also its distribution over the hide, which is in-

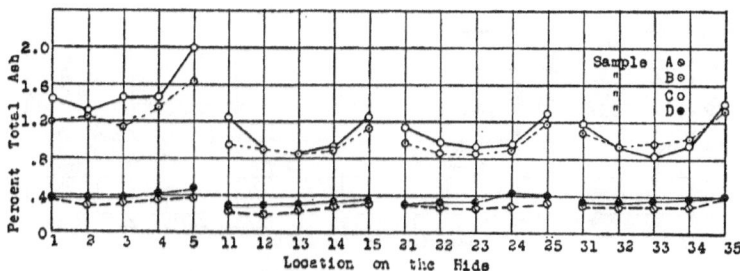

Fig. 11.—*Chart showing variation of total ash over the hide*

fluenced by much the same factors as is the distribution of the other water solubles. The variations on the bend are small where the total glucose is low, but increase considerably when the content is larger, as in sample A.

Total Ash and Epsom Salts.—All vegetable tanned leather will naturally contain a small amount of ash, most of which is derived from the residual lime left from the beam-house operations, but where a high ash is found it is generally caused by the presence of added salts. The total ash is influenced more by the amount of Epsom salts added to the leather. The salts are generally added with the glucose and show (Fig. 10) the same variation over the hide as the glucose and water solubles, with the exception of samples A and D. The uniform curve and small amount found

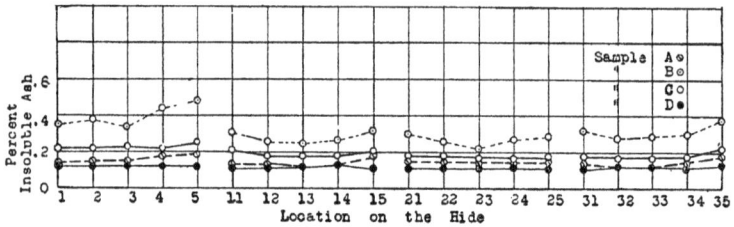

Fig. 12.—*Chart showing variation of insoluble ash over the hide*

in all locations of sample A indicate that only that naturally found in the leather was present, while the generally uniform distribution for sample D shows that little salts was added. The curves for the total and insoluble ash (Figs. 11 and 12) therefore naturally follow the same general trend as those for the salts (Fig. 10). The analyses show that the salts are practically constant for all locations where none is added, but that there is an

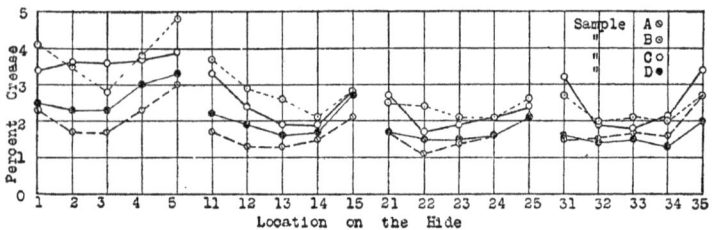

Fig. 13.—*Chart showing distribution of grease over the hide*

unequal absorption when only small amounts are added and a greater variation when larger amounts are used.

Petroleum Ether Extract (Fats and Oils).—A small amount of fats and oils occur naturally in leather, and it is customary to add some in the manufacturing processes to prevent the grain from becoming harsh and cracky, to properly lubricate the fibers, and to prevent too rapid drying, which produces a light-colored leather.

Fig. 13 shows that the grease content does not vary much for the four brands but does vary considerably over the bend, the more open portions showing a higher grease content.

Hide Substance.—The amount of hide substance is directly dependent upon the amount of materials added to the leather in

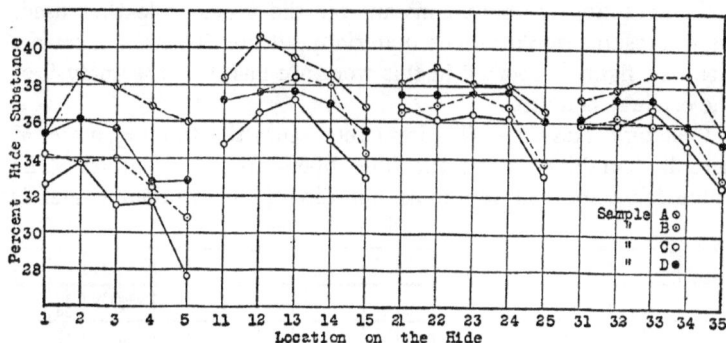

FIG. 14.—*Chart showing distribution of hide substance over the hide*

the process of manufacture. The raw hide is nearly all hide substance. A high value would indicate a lightly tanned and unloaded leather and a very low value a heavily tanned or highly loaded leather, which facts make it logical to assume that the

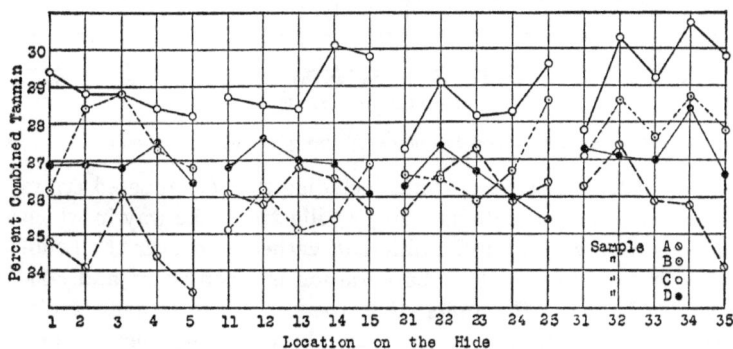

FIG. 15.—*Chart showing distribution of combined tannin over the hide*

percentage of hide substance would be high or low according to whether the water solubles and combined tannin were low or high respectively. This is generally the case, as will be noticed by comparing Fig. 14 with those for the water solubles and combined tannin.

There is always an error in the amount of hide substance, as determined on the original sample, corresponding to the amount of nitrogen in the water extract and the petroleum ether extract. The latter is undoubtedly negligible, and the former was found to be the equivalent of about 0.2 to 0.3 per cent of the leather. This correction is about constant for the types of leather used. As the results would not be materially affected by such a correction, the figures reported in this work are based on the analysis of the original samples.

Combined Tannin.—This item represents the tanning material actually combined with the hide fibers to form leather. The value for combined tannin is not directly determinable and is

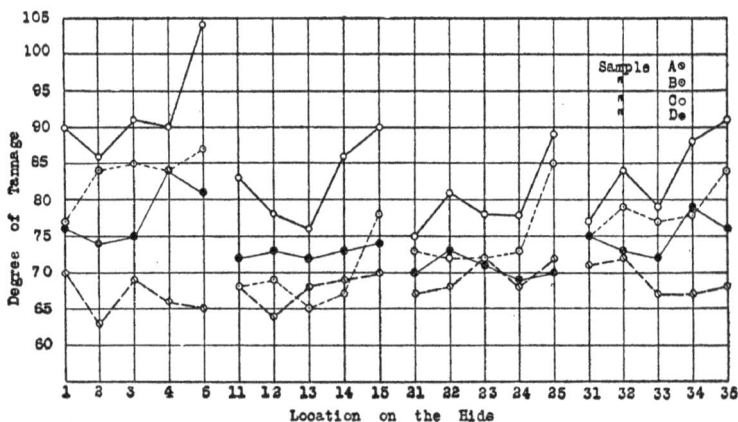

FIG. 16.—*Chart showing variation of degree of tannage over the hide*

arrived at by "difference," thus including the accumulated errors in the other determinations. Fig. 15 illustrates the way in which this quantity varies for the different leathers and over the bend. While it would seem that the variation is inconsistent and considerable between the different leathers, there is actually a difference of only about 3 per cent between the highest and lowest average values.

Degree of Tannage.—The value for the degree of tannage shows the number of parts of tannin combined with 100 parts of hide substance and is affected by the other determinations only in so far as their errors are contained in the values for the combined tannin. Fig. 16 shows that the value varies considerably between the different leathers and also over the hide. The greatest variation for one sample of leather occurs in the case of sample C, which was given a further tanning in drums after the preliminary tannage in the vats. The average degree of tannage for the different leathers varied from 68 to 84.

Ratio of Tannins to Nontannins.—Fig. 17 shows the variation in the ratio of the tannins to the nontannins. This variation is due largely to the presence of the glucose and salts added to the leathers, since the ratio of tannins to nontannins other than these materials varies only from 2.2 to 2.8. The graph shows that the values of the ratios for the four brands of leather are widely separated and afford an excellent indication as to the nature of the water-soluble material in the leathers. The effect of added glucose on this ratio is shown in the case of sample A, which has the

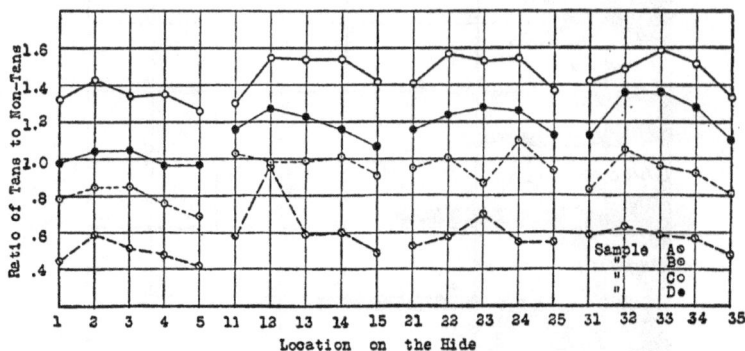

FIG. 17.—*Chart showing variation in ratio of tannins to nontannins over the hide*

highest glucose content and the lowest ratio of tannins to nontannins. This indicates that the water soluble is made up to a great extent of soluble nontannin in contrast with sample C, which has the highest value of the ratio of tannins to nontannins, which shows that the water solubles are made up of a greater percentage of soluble tannins.

(c) ANALYSES OF WORN SOLES.—The results of the analyses of worn soles are presented numerically in Table 2. A discussion of these analyses as compared with the analyses of the original leather is embodied later under heading IV-3-(c) (p. 34).

(d) SPECIFIC GRAVITY TESTS.—The specific gravity is the weight of a substance compared with the weight of an equal volume of water. Thus by the determination of this value for the different leathers information as to their relative densities may be obtained. Fig. 19 shows the values of the specific gravity for each

Insoluble Ash
0   2   4

Each Block Arranged
In the following Order
Sample C
"   A
"   D
"   B

Total Ash
0   4   8   12

Water Soluble
18   20   22

Petroleum Ether Extract
0   1   2   3

Epsom Salts
0   5   1.0   1.5   2.0

Glucose
0   2   4   6   8

Degree of Tannage
55   60   65   70   75   80

Combined Tannin
23   24   25   26   27   28   29

Hide Substance
26   28   30   32   34   36   38

Ratio of Tans to Non-Tans
0   2   4   .6   8   1.0   1.2   1.4   1.6

Wear in Days Per Iron
0   1   2   3   4   5   6   7

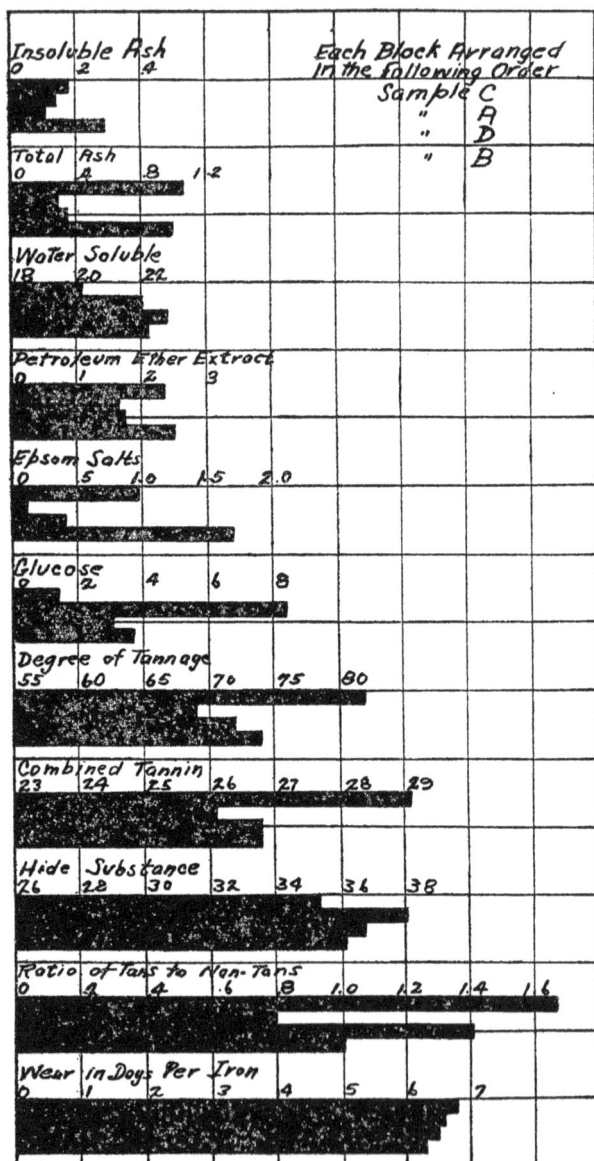

FIG. 18.—*Chart showing average values for the chemical constituents of
the original leather and the average wear*

brand of leather for each location on the hide. There is a general tendency for the density to decrease in value as the location varies from the backbone edge to the belly edge. The variation

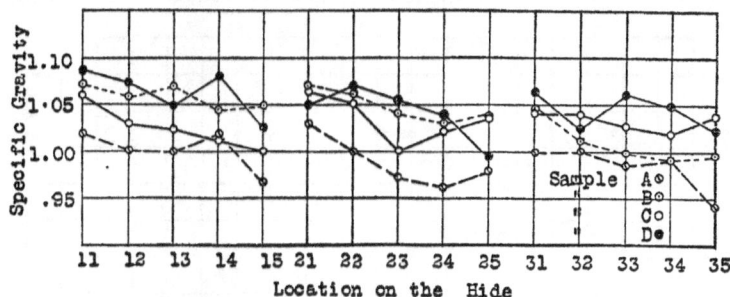

FIG. 19.—*Chart showing variation in density over the hide*

in density over the hide for any particular leather is not greater than 9 per cent.

On an equal volume basis, sample A was the lightest in weight. This is surprising, since it contained an average glucose content of 8.8 per cent. This leather had the lowest value for the degree of tannage, which value, however, was high enough to indicate a well-tanned leather, and also had the lowest percentage of soluble

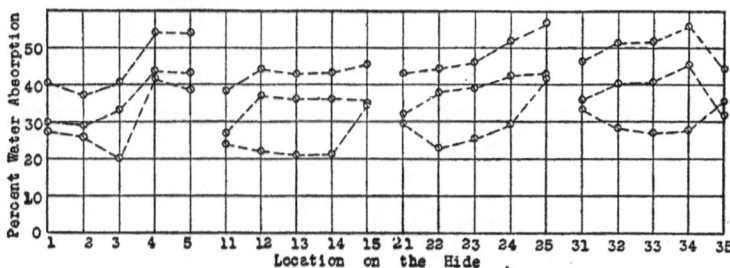

FIG. 20.—*Chart showing the variation in water absorption over the hide for sample A*

tanning material. The addition of a large quantity of glucose was not sufficient to make the weight equal to that of the other leathers with a higher value for the degree of tannage. Sample D, which was a leather well tanned by the layaway process, had the highest gravity.

(e) WATER ABSORPTION TESTS.—Figs. 20 to 23, inclusive, show graphically the variation in water absorption over the hide for the different leathers. There is a general tendency for the more spongy parts to absorb a relatively greater amount. The lower curve on each chart represents the water absorption in 30 minutes. The amount of soluble matter soaked out in the 30-minute test was negligible, so that no correction was made. The middle

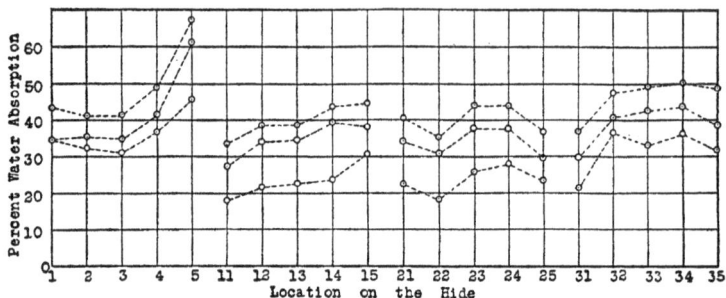

FIG. 21.—*Chart showing the variation in water absorption over the hide for sample B*

curve represents the apparent water absorption in 24 hours, while the upper curve represents the actual absorption. The values for the upper curve were obtained by correcting the apparent absorption values by adding the percentage of soluble materials soaked from the samples in the 24-hour test. The amounts lost into solution are represented by the difference be-

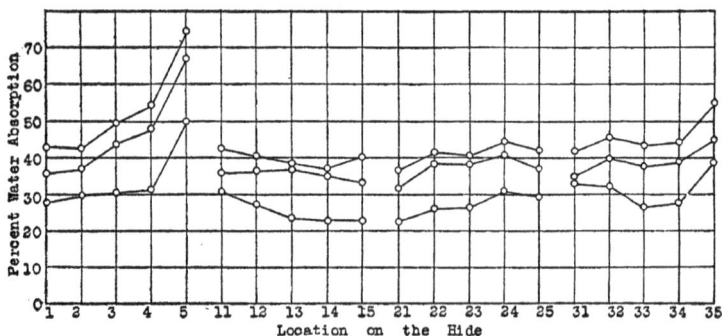

FIG. 22.—*Chart showing variation in water absorption over the hide for sample C*

tween the upper and middle curves and amount in some cases to as much as 15 per cent. There is a general parallelism between the curves for the 30-minute and 24-hour tests, and hence it would appear that a 30-minute test for water absorption qualities would be practicable, since no correction needs to be made for the loss of soluble material into solution.

The loss of dry matter in the 24-hour test seems to be influenced by the water solubles, especially the glucose, since the relative loss of glucose during the test is greater than that for any other material. This statement is substantiated by the analysis of the worn soles, which shows that in actual wear the leather with the highest glucose content lost the greatest amount of water solubles.

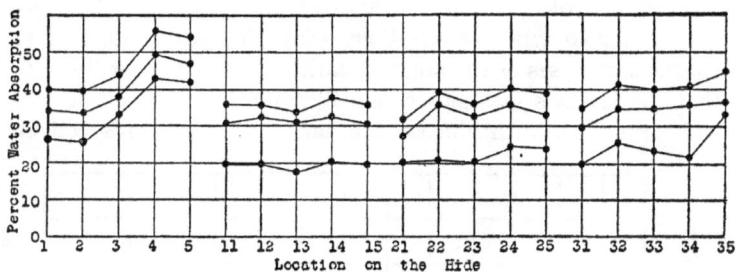

FIG. 23.—*Chart showing the variation in water absorption over the hide for sample D*

Table 3 gives the average values for the water absorption of the different leathers.

TABLE 3.—Water Absorption Tests

Sample	Percentage absorption in 30 minutes	Percentage apparent absorption in 24 hours	Percentage soaked out in 24 hours	Percentage actual absorption in 24 hours
A.	29.03	36.78	9.99	46.77
B.	28.66	37.33	6.38	43.70
C.	28.63	39.56	5.36	44.92
D.	24.95	34.66	5.25	39.91

An interesting point regarding these figures is the fact that the percentage of material soaked out in the 24-hour test was greater for samples A and B, which leathers contain large amounts of added glucose or salts.

3. COMPARISON OF FIELD AND LABORATORY TESTS

(a) MACHINE AND FIELD WEARING TESTS.—The upper curve of Fig. 24 represents the average wear in days per iron for all the soles of the four brands of leather. The lower curve represents the wear, as indicated by the machine tests, expressed as the average loss in volume of the test pieces as compared with a standard specimen of composition material, a sample of which was tested with each group of six samples of leather. The two curves show the same general tendency as regards wearing quality

with the exception of location No. 31.   No cause can be assigned
for this apparent discrepancy, but it is expected that subsequent
investigations will correct this difference.   Although the wearing
machine is still in the experimental stage, the results obtained
with it are sufficiently consistent to give an indication of the
wearing quality of several samples of the same kind of leather.

(*b*) Relation Between the Composition of the Original
Leather and the Wear.—Comparing the wear data and the
chemical analyses of the original leathers (Fig. 18), it will be seen
that the leathers (samples A and C) tanned to the belting stage
and then filled with glucose and tanning material, respectively,

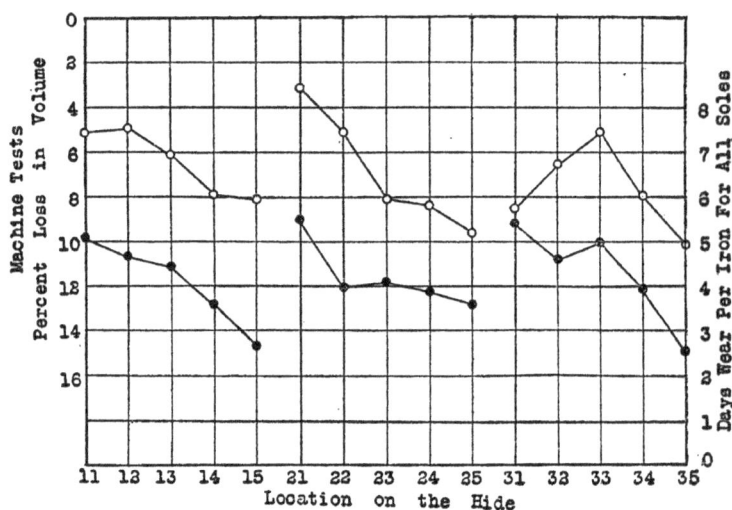

Fig. 24.—*Chart showing relation between actual service tests and laboratory machine tests*

by drumming, gave the longest wear, but the difference was not
large.   Of these two filling materials the glucose in sample A
was practically all lost during wear while the tanning material
in sample C was not lost to any great extent.

(*c*) Comparison of the Analyses of the Original Leathers
and Those of the Worn Soles.—The data tabulated in Table
2 are the averages of analyses made on all the soles that were
tested and of the corresponding blocks before test.   In cases
where no worn soles from a given block were available for analysis,
the data for the original block were not included in the average.
(Soles Nos. 1 to 5.)   The data are therefore strictly comparable,
and the number of blocks included in the averages given are noted
in the table.

Glucose.—The loss of all the added glucose is clearly shown by the values in Table 2. The loss is greatest for sample A, which contained the highest percentage of added glucose.

Epsom Salts.—The amount of magnesium, calculated as Epsom salts, appears to come to a dead level of about 0.7 per cent in the worn soles. (Table 2.) There is evidently a loss of magnesium where the original material contained much salts and also an accumulation of magnesium during wear, as shown by the analysis of sample A soles which had considerably more magnesium in the worn soles than in the original leather. This accumulation is probably due to the contact of magnesium-containing matter with the outer surface of the sole. It is quite likely that all the magnesium present as Epsom salts was leached out of the leather during wear, and that the magnesium finally present was largely derived from mechanical inclusions.

Total Ash.—The total ash is greatly increased during wear. This increase is largely insoluble material, but considerable soluble mineral matter is included also.

Insoluble Ash.—The increase in insoluble ash is undoubtedly due to the mechanical acquisition of mineral matter, which can be shown by examination of the surface of any leather sole from a worn shoe.

Petroleum Ether Extract.—There is an increase in the grease content of all the soles tested, amounting to about 2 per cent on an average. As no dubbing was used, this shows a tendency on the part of the sole to absorb any grease with which it comes in contact. Oiled floors might account for some of the increase, but it is possible that the grease becomes concentrated in the remaining portions of the sole as the leather is worn away. The absorption of ether-soluble material from the foot is a doubtful possibility on account of the construction of the shoes. The extracted grease, which was much darker than the grease from the original blocks, indicated that the composition of the grease was considerably different than the original extract. The iodine numbers of the greases from the original leathers were 53.5, 42.1, 40.1, and 72.6, respectively, for A, B, C, and D. In the same order, the iodine numbers of the greases extracted from the worn soles were 55.6, 42.8, 44.2, and 48.5. These data show that in only one tannage was there any material change in the iodine number. The original oil used on this tannage contained a high percentage of cod oil, and the decrease noted is easily accounted for by the oxidation of the oil in the leather. The iodine numbers of the oils from the other tannages indicate that less drying oil was used, but it is rather hard to account for the fact that there was no decrease at all in the iodine numbers of these oils.

Water-Soluble Material.—Except for the loss of glucose and Epsom salts (Table 2), there appears to be very little loss of water-soluble material from the soles. This is probably influenced to some extent by the fact that much of the material soluble at 50° C. is insoluble at ordinary temperatures and would not be lost from the leather.

Hide Substance.—There was practically no change in the percentage of hide substance found in the leather before and after wear. The loss of glucose and Epsom salts seemed to be about compensated for by the increase in total ash and grease.

Combined Tannin.—The relative changes in this figure were slight in all cases, which shows that this material is firmly fixed in the leather, and is not lost to any extent during wear.

Degree of Tannage.—When consideration is taken of the wide variation of this ratio in the original leather there was relatively little change during wear. The extraction of water-soluble material at 50° C. is a rather severe test, and a tannage stable at that temperature would not be likely to be affected by conditions experienced during ordinary wear.

The Ratio of Tannins to Nontannins in the Water-Soluble Material.—This ratio is not affected as much as would be expected from the fact that there was a loss of glucose and Epsom salts from the leather. It must be that some of the tannins are leached out and replaced by nontannins; or that there is an actual change in the character of the material which, when extracted from the original leather, was absorbed by hide powder. There was some increase in this ratio in the case of A, which lost considerable glucose during wear, but not enough to account for the loss of glucose. In the other tannages there seemed to be an actual decrease in this ratio. No explanation of these facts is offered at this time.

Summary.—The effects of wear on the chemical composition of the leather may be summarized as follows:

1. Under the conditions of the test the greater portion of the added glucose and salts was lost during wear, but no great decrease in the other constituents of the water soluble was apparent.

2. The leather substance was not affected appreciably during wear.

3. There was an actual increase in the grease content, for which no definite explanation can be offered.

4. There was a large increase in the ash content of the leather, largely owing to mechanical accretion.

(*d*) SPECIFIC GRAVITY AND FIELD WEARING TESTS.—Table 4 shows the days wear per iron, the average specific gravity, and the days wear per iron per unit of weight.

TABLE 4

Sample	Days wear per iron	Average values, specific gravity	Days wear (per iron) per unit of weight
A.	6.6	0.992	6.65
B.	6.3	1.041	6.05
C.	6.8	1.033	6.58
D.	6.5	1.050	6.20

From these values it will be seen that sample A, which was second in wearing quality, actually becomes first in wearing quality when compared on a basis of unit weight. Thus it would appear that a leather with low density or light weight could be purchased, and while a lower value for wearing in days per iron would be obtained as compared with heavier leathers, a wear per unit of weight might be secured which would render it much cheaper leather on a cost basis.

(*e*) COMPARISON OF THE WEAR DATA WITH THE WATER ABSORPTION.—Comparing the wear data with the water absorption, Figs. 6, 20, 21, 22, and 23, it appears that the greater the absorption the poorer the leather. This shows that the same factors that tend to cause a high water absorption also tend to decrease the life of the leather. That this is entirely due to the action of the water is improbable, as it would not be likely that a belly sole would wear as well as one from the back, even if they were both protected from the action of water by suitable stuffing. It is probable, however, that if a bend were heavily stuffed with wax there would be less difference shown between the wear of a back and a belly sole than is shown by the unstuffed leathers used in these experiments.

V. CONCLUSIONS

From the results of this investigation it would appear that the four brands of leather tested did not differ greatly in wearing quality. There is no indication that the addition of glucose and salts is either beneficial or detrimental to the durability of the leather. It is shown conclusively that the greater part of the added glucose and salts is lost·from the leather during wear, while the other water-soluble materials appear to be retained in the leather. It is also shown that the leathers A and C, which were given the same tanning in the layaways and then filled with glucose and tanning material, respectively, by drumming, have the same wearing quality. The method of adding the tanning materials, either by drumming (sample A) or of giving a long-time tanning in the layaways (sample D), also appeared to have little effect on the wearing quality. When further tests are completed it is expected that more definite and conclusive results will be secured to show the effects of glucose and salts on the wearing quality of sole leather.

WASHINGTON, April 26, 1919.